S0-BXV-023

**Grade K
Module 2**

Eureka Math™
A Story of Units

Special thanks go to the Gordan A. Cain Center and to the Department of Mathematics at Louisiana State University for their support in the development of *Eureka Math*.

Published by Common Core

Printed in the U.S.A.
This book may be purchased from the publisher at commoncore.org
10 9 8 7 6 5 4 3 2 1
ISBN 978-1-63255-001-9

Name _____   Date _____

Sort the shapes.

Shapes with a Curve	Shapes without a Curve

    Lesson 1:   Find and describe flat triangles, square, rectangles, hexagons, and
circles using informal language without naming.

1

Lesson 1:   Find and describe flat triangles, square, rectangles, hexagons, and circles using informal language without naming.

Name _____   Date _____

Draw a line from the shape to its matching object.

EUREKA MATH™ | Lesson 1: Find and describe flat triangles, square, rectangles, hexagons, and circles using informal language without naming.

3

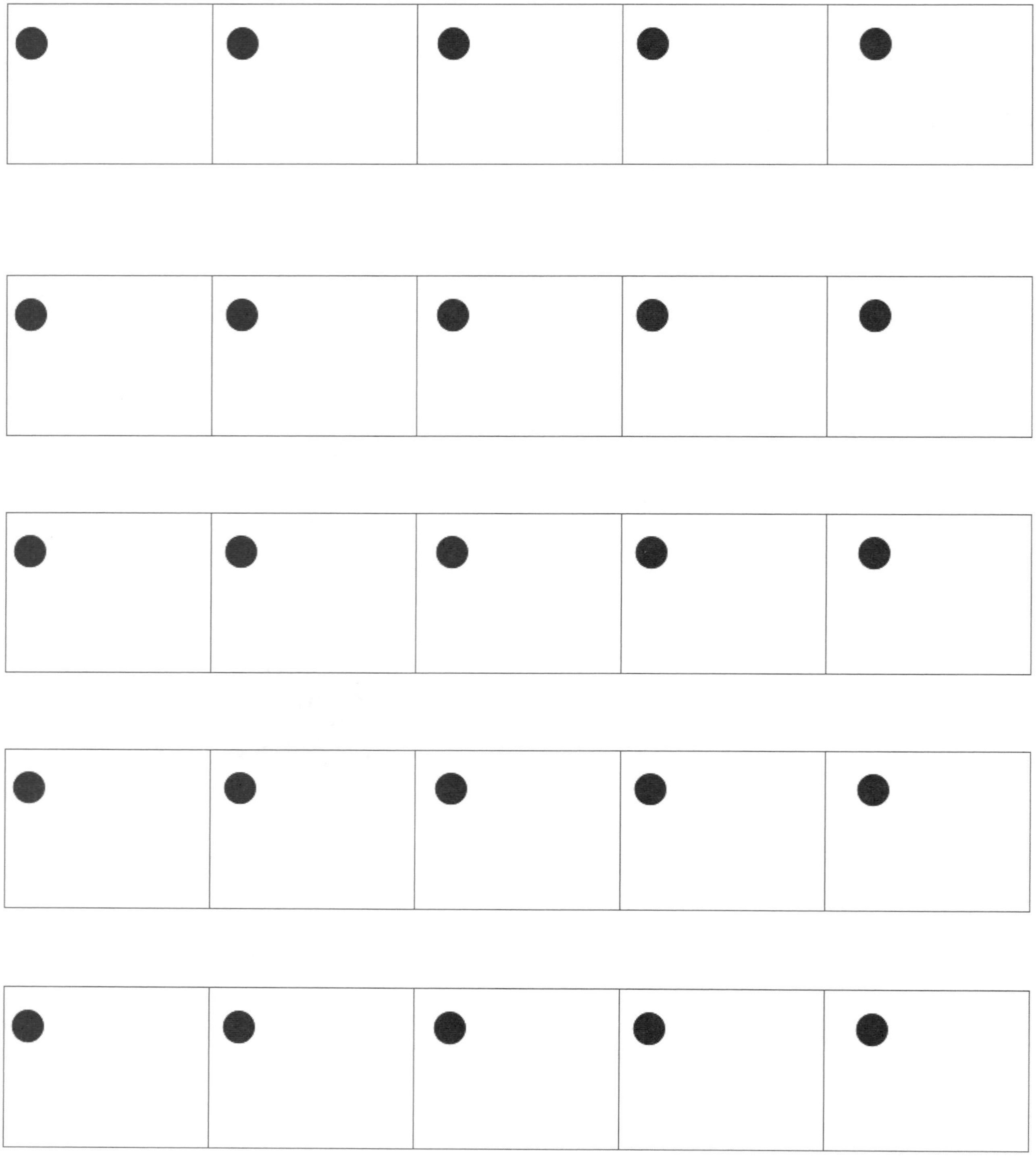

5-group mat

EUREKA MATH

Lesson 1: Find and describe flat triangles, square, rectangles, hexagons, and circles using informal language without naming.

4

Draw more to make 5.

draw more

EUREKA
MATH

Lesson 1:   Find and describe flat triangles, square, rectangles, hexagons, and circles using informal language without naming.

5

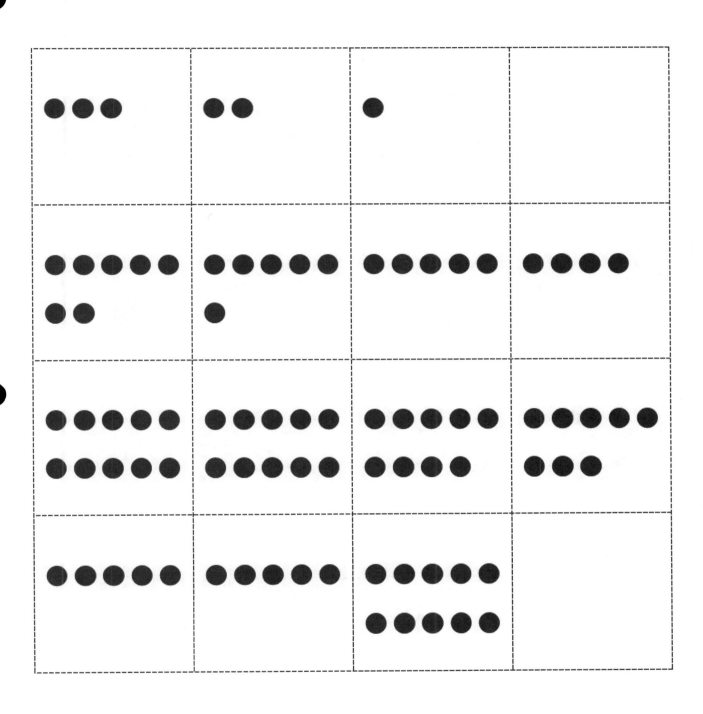

5-group cards

EUREKA
MATH™

Lesson 1:   Find and describe flat triangles, square, rectangles, hexagons, and
             circles using informal language without naming.

7

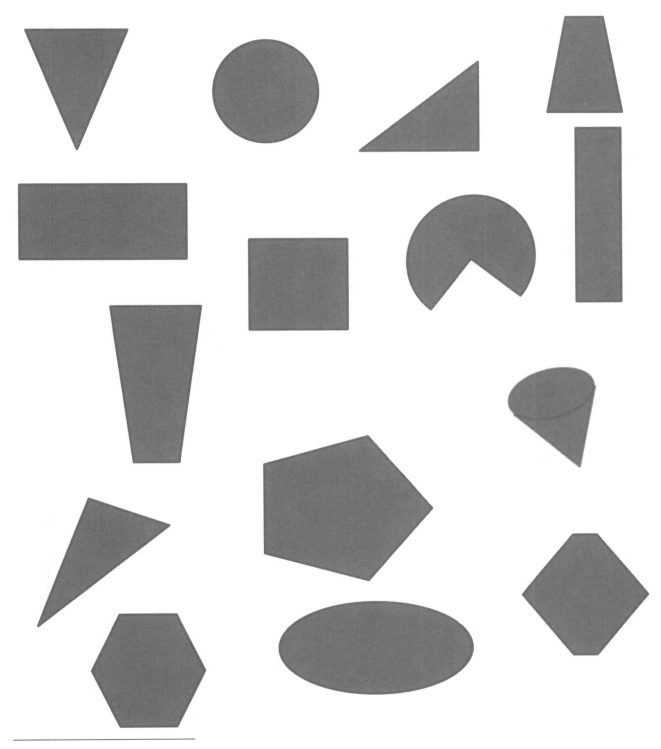

shapes

Lesson 1: Find and describe flat triangles, square, rectangles, hexagons, and circles using informal language without naming.

9

Name _____   Date _____

Find the triangles and color them blue. Put an X on shapes that are not triangles.

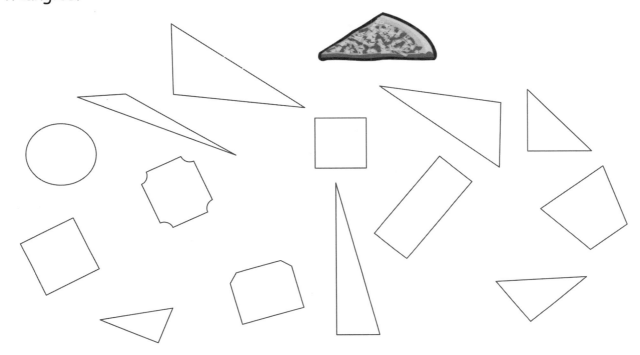

Draw some triangles.

EUREKA
MATH™   Lesson 2:   Explain decisions about classifications of triangles into categories using
                           variants and non examples. Identify shapes as triangles.

11

Name _____   Date _____

Color the triangles red and the other shapes blue.

Draw 2 different triangles of your own.

Lesson 2:   Explain decisions about classifications of triangles into categories using
              variants and non examples. Identify shapes as triangles.

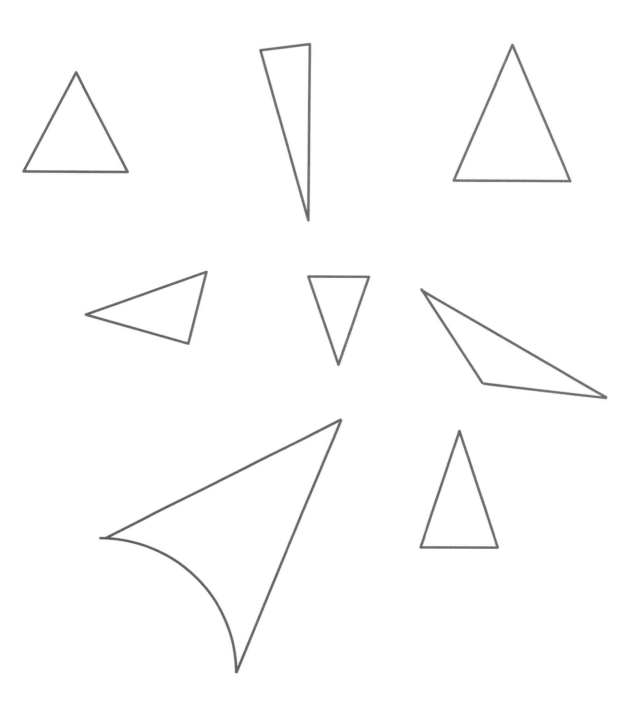

shapes

EUREKA MATH™ | **Lesson 2:**   Explain decisions about classifications of triangles into categories using variants and non examples. Identify shapes as triangles.

Name _____   Date _____

Find the rectangles and color them red. Put an X on shapes that are not rectangles.

Draw some rectangles.

   Lesson 3:   Explain decisions about classifications of rectangles into categories using variants and non examples. Identify shapes as rectangles

15

Name _____   Date _____

Color all the rectangles red. Color all the triangles green.

On the back of your paper, draw 2 rectangles and 3 triangles.

How many shapes did you draw? Put your answer in the circle.

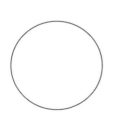

EUREKA
MATH™

Lesson 3:   Explain decisions about classifications of rectangles into categories
            using variants and non examples. Identify shapes as rectangles

16

shapes

EUREKA MATH

| Lesson 3: | Explain decisions about classifications of rectangles into categories using variants and non examples. Identify shapes as rectangles

17

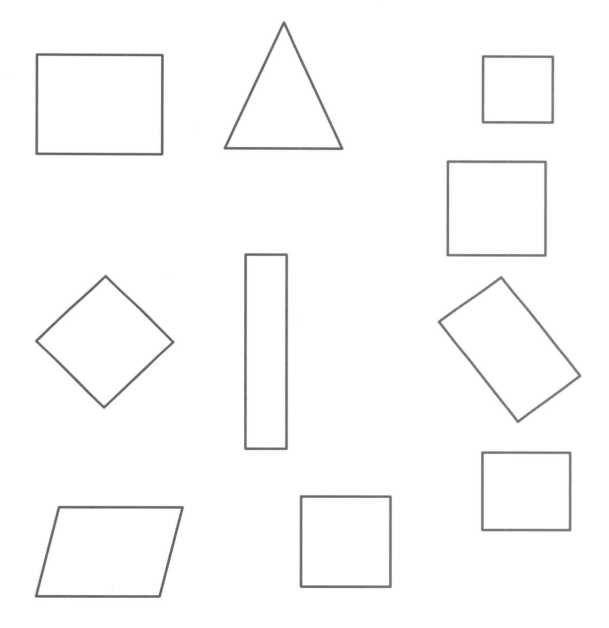

shapes

EUREKA MATH™

Lesson 3:   Explain decisions about classifications of rectangles into categories
using variants and non examples. Identify shapes as rectangles

19

dot paper

Name _____   Date _____

Find the circles and color them green. Find the hexagons and color them yellow. Put an X on shapes that are not hexagons or circles.

Draw hexagons and circles.

EUREKA MATH | Lesson 4:   Explain decisions about classifications of hexagons and circles, and identify them by name. Make observations using variants and non examples.

23

Name_____ Date_____

Color the triangles blue.

Color the rectangles red.

Color the circles green.

Color the hexagons yellow.

On the back of your paper draw 2 triangles and 1 hexagon.

How many shapes did you draw? _____

EUREKA
MATH™

Lesson 4: Explain decisions about classifications of hexagons and circles, and
identify them by name. Make observations using variants and non
examples.

24

shapes

Lesson 4:   Explain decisions about classifications of hexagons and circles, and
identify them by name. Make observations using variants and non
examples.

25

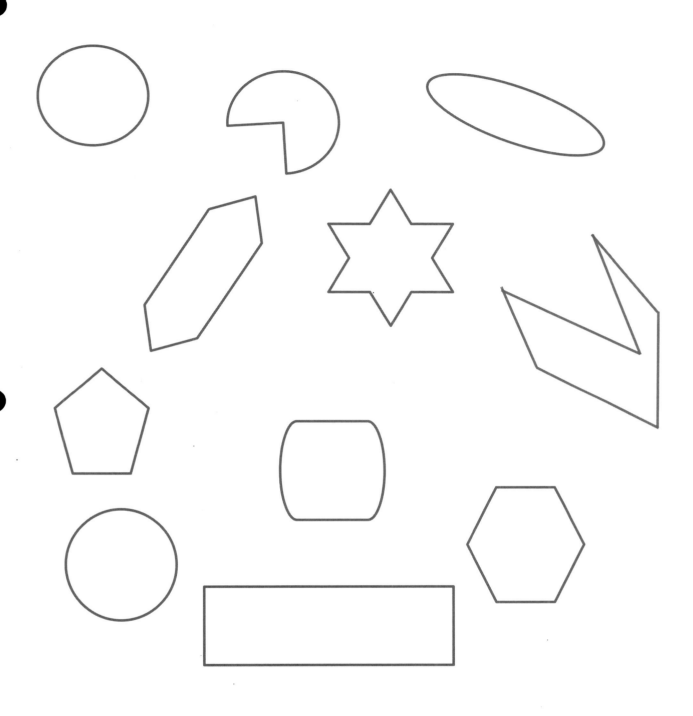

shapes

EUREKA
MATH™

Lesson 4:   Explain decisions about classifications of hexagons and circles, and
identify them by name. Make observations using variants and non
examples.

27

Name _____   Date _____

Cut out all of the shapes, and put them next to your paper with the duck.
Listen to the directions, and glue the objects onto your paper.

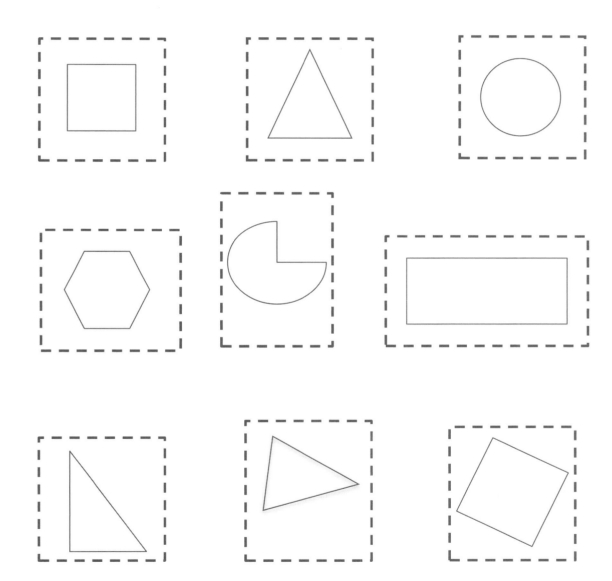

EUREKA
MATH™

Lesson 5:   Describe and communicate positions of all flat shapes using the words
            above, below, beside, in front of, next to, and *behind*

29

Name _____   Date _____

EUREKA MATH | Lesson 5:   Describe and communicate positions of all flat shapes using the words
                                      above, below, beside, in front of, next to, and *behind*

31

Name _____   Date _____

- **Behind** the elephant, draw a shape with 4 straight sides that are exactly the same length. Color it blue.

- **Above** the elephant, draw a shape with no corners. Color it yellow.

- **In front of** the elephant, draw a shape with 3 straight sides. Color it green.

- **Below** the elephant, draw a shape with 4 sides, 2 long and 2 short. Color it red.

- **Below** the elephant, draw a shape with 6 corners. Color it orange.

On the back of your paper draw 1 hexagon and 4 triangles.
How many shapes did you draw? Put your answer in the circle.

Lesson 5:   Describe and communicate positions of all flat shapes using the words
              above, below, beside, in front of, next to, and *behind*

32

triangle

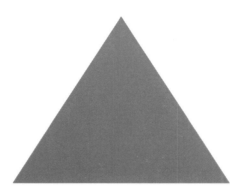

rectangle

_____

signs

Lesson 5:   Describe and communicate positions of all flat shapes using the words
above, below, beside, in front of, next to, and *behind*

33

square

hexagon

signs

EUREKA MATH™ | Lesson 5: Describe and communicate positions of all flat shapes using the words *above, below, beside, in front of, next to,* and *behind*

35

circle

signs

EUREKA MATH

Lesson 5: Describe and communicate positions of all flat shapes using the words *above, below, beside, in front of, next to,* and *behind*

37

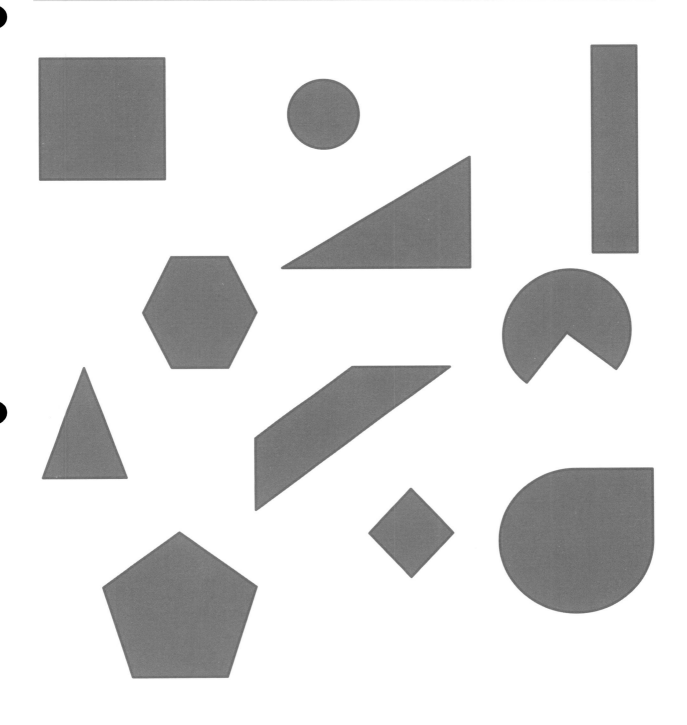

shapes

EUREKA MATH

Lesson 5:   Describe and communicate positions of all flat shapes using the words
above, below, beside, in front of, next to, and *behind*

39

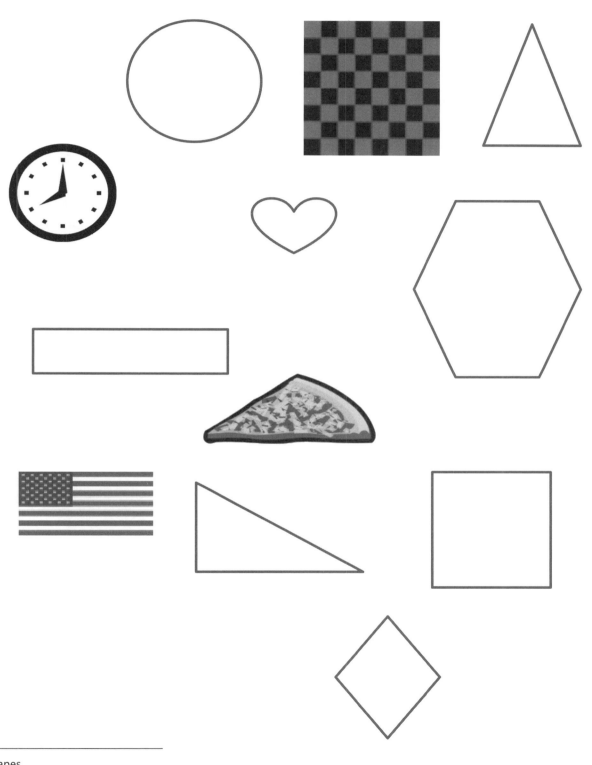

shapes

Lesson 5: Describe and communicate positions of all flat shapes using the words *above, below, beside, in front of, next to,* and *behind*

41

shapes

EUREKA
MATH™

Lesson 5:   Describe and communicate positions of all flat shapes using the words
above, below, beside, in front of, next to, and *behind*

43

Name _____   Date _____

Match these objects and solids by drawing a line with your ruler from the object to the solid.

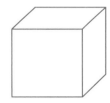

On the back of the paper, draw solid shapes that you see in the classroom.

EUREKA MATH™ | Lesson 6:   Find and describe solid shapes using informal language without naming.

45

Name _____ Date _____

Find things in your house or in a magazine that look like these solids. Draw the solids or cut out and paste pictures from a magazine.

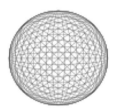

EUREKA
MATH™

Lesson 6: Find and describe solid shapes using informal language without naming.

46

| 1 | 2 | 3 | 4 | 5 | 6 | 7 | 8 | 9 | 10 |

| 1 | 2 | 3 | 4 | 5 | 6 | 7 | 8 | 9 | 10 |

| 1 | 2 | 3 | 4 | 5 | 6 | 7 | 8 | 9 | 10 |

| 1 | 2 | 3 | 4 | 5 | 6 | 7 | 8 | 9 | 10 |

number path

Lesson 6:   Find and describe solid shapes using informal language without naming.

Name _____   Date _____

Circle the cylinders with red.

Circle the cubes with yellow.

Circle the cones with green.

Circle the spheres with blue.

EUREKA
MATH™

Lesson 7:   Explain decisions about classification of solid shapes into categories.
Name the solid shapes.

49

Name _____   Date _____

Cut one set of solid shapes. Sort the 4 solid shapes. Paste onto the chart.

These have corners.	These do not have corners.

Cut the other set of solid shapes and make a rule for your sort. Paste onto the chart.

Lesson 7:   Explain decisions about classification of solid shapes into categories.
            Name the solid shapes.

50

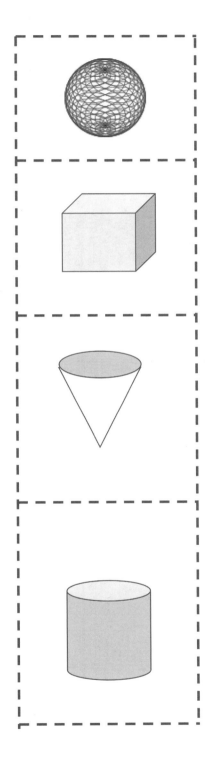

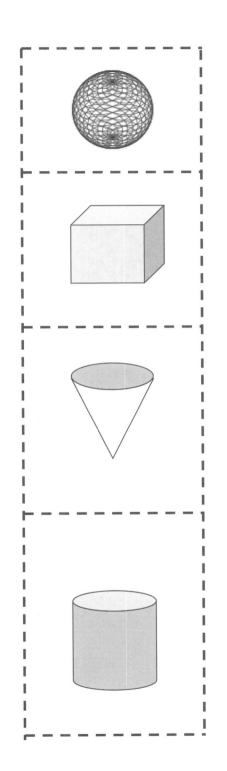

EUREKA
MATH

Lesson 7:   Explain decisions about classification of solid shapes into categories.
Name the solid shapes.

51

Name _____   Date _____

EUREKA MATH | Lesson 8:   Describe and communicate positions of all solid shapes using the
                                words *above, below, beside, in front of, next to,* and *behind*

53

Name _____   Date _____

Directions: Read to students.

Paste the sphere **above** the train.
Paste the cube **behind** the train.
Paste the cylinder **in front of** the train.
Paste the cone **below** the train.

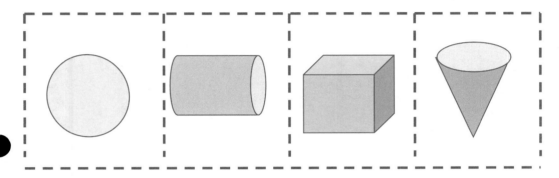

Provide one strip for every student.

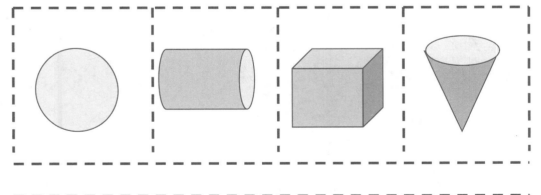

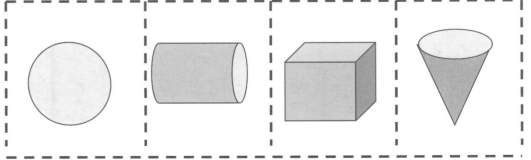

EUREKA MATH

Lesson 8:   Describe and communicate positions of all solid shapes using the
words *above, below, beside, in front of, next to,* and *behind*

55

Name _____   Date _____

Tell someone at home the names of each solid shape.

   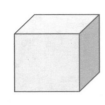

Sphere   Cylinder   Cone   Cube

Color the car **beside** the stop sign green.

Circle the **next** car with blue.

Color the car **behind** the circled car red.

Draw a road **below** the cars.

Draw a policeman **in front of** the cars.

Draw a sun **above** the cars.

EUREKA
MATH™

Lesson 8:   Describe and communicate positions of all solid shapes using the
             words *above, below, beside, in front of, next to,* and *behind*

57

Name _____   Date _____

Circle the pictures of the flat shapes with red. Circle the pictures of the solid shapes with green.

Name _____ Date _____

In each row, circle the one that doesn't belong. Explain your choice to a grown-up.

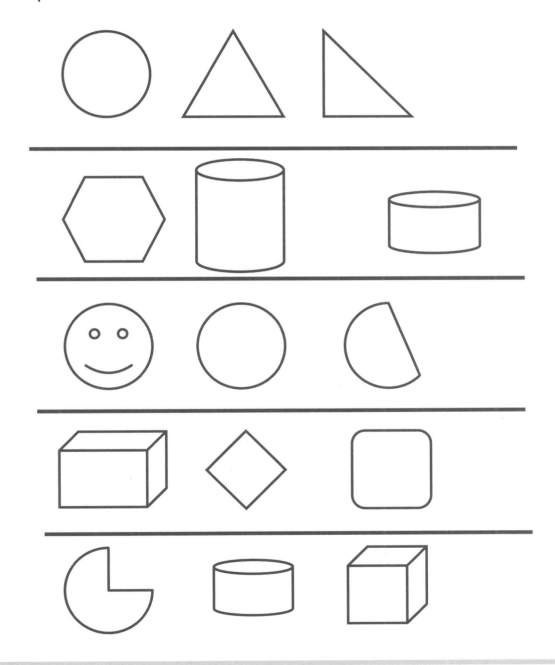

Lesson 9:   Identify and sort shapes as two-dimensional or three-dimensional,
recognize two-dimensional and three dimensional shapes in different
orientation and sizes.

59

Name _____   Date _____

Shape Up Your Kitchen!

Search your kitchen to see what shapes and solids you can find. Make a kitchen-shaped collage by drawing the shapes that you see and by tracing the faces of the solids that you find. Color your collage.

Lesson 10:   Culminating task—collaborative groups create displays of different flat shapes with examples, non examples, and a corresponding solid shape.

61

Name _____   Date _____

These are (____).   These are not (____).

work mat

EUREKA
MATH™

| Lesson 10: | Culminating task—collaborative groups create displays of different flat shapes with examples, non examples, and a corresponding solid shape.

62

Student Name _____

Topic A: Two-Dimensional Flat Shapes

Rubric Score: _____ Time Elapsed: _____

Materials: (S) Paper cutouts of typical triangles, squares, rectangles, hexagons, and circles; paper cutouts of variant shapes and difficult distracters (see Geometry Progression, p.6)

	Date 1	Date 2	Date 3
Topic A			
Topic B			
Topic C			

1. (Hold up a rectangle. Use different shapes for each student.) Point to something in this room that is the same shape and use your words to tell me all about it. How do you know they are the same shape?

2. (Place several typical, variant, and distracting shapes on the desk. Be sure to include three or four triangles.) Please put all the triangles in my hand. How can you tell they were all triangles?

3. (Hold up a rectangle.) How is a triangle different from this rectangle? How is it the same?

4. (Place five typical shapes in front of the student.) Put the circle next to the rectangle. Put the square below the hexagon. Put the triangle beside the square.

What did the student do?	What did the student say?
1.	
2.	
3.	
4.	

Topic B: Three-Dimensional Solid Shapes

Rubric Score: _____   Time Elapsed: _____

Materials: 1 cone; 3 cylinders (wooden or plastic); a variety of real solid shapes, e.g., soup can, paper towel
roll, party hat, ball, dice, or an unsharpened cylindrical (not hexagonal prism) pencil

1. (Hand a cylinder to the student.) Point to something in this room that is the same solid shape, and use your words to tell me all about it.

2. (Place seven solid shapes in front of the student including three cylinders: wooden, plastic, realia.) Put all the cylinders in this box.

3. (Show a cone.) How is the cylinder you are holding different from this cone? How is it the same?

4. (Place the set of solid shapes in front of the student.) Put the cube in front of the cylinder. Put the sphere behind the cone. Put the cone above the cube.

What did the student do?	What did the student say?
1.	
2.	
3.	
4.	

Topic C: Two-Dimensional and Three-Dimensional Shapes

Rubric Score: _____ Time Elapsed: _____

Materials: Set of flat and solid shapes (do not use the paper cutouts from Topic A, but rather both commercial flat shapes and classroom flat shapes, such as a piece of colored construction paper, a CD sleeve, or a name tag)

1. Can you sort these shapes into one group of flat shapes and one group of solid shapes?
2. Tell me about your groups. What is the same about both groups? What is different?
3. Can you sort these shapes a different way? Tell me about your new groups. What is the same? What is different?

What did the student do?	What did the student say?
1.	
2.	
3.	

Notes

Notes